C000148107

PRUNING FC
BEGINNERS

:The comprehensive guide on how,when and best season to prune your plant

Larry Pat

Copyright © by Larry Pat 2023. All rights reserved.

Before this document is duplicated or reproduced in any manner, the publisher's consent must be gained. Therefore, the contents within can neither be stored electronically, transferred, nor kept in a database. Neither in Part nor full can the document be copied, scanned, faxed, or retained without approval from the publisher or creator.

TABLE OF CONTENT

INTRODUCTION

Pruning is a fundamental consideration step for developing natural product trees; it's not discretionary. Accepting that you never need to prune is the principal pruning botch that grounds-keepers make, and they keep on making more as they endeavor to prune.

Pruning is a critical stage under the watchful eye of your natural product trees. It supports development in the spring, opens the branches to make more air dissemination, permits more daylight to arrive at the leaves, and makes a by and large better natural product tree.

One of the most intimidating tasks for a new Gardener while really focusing on any plant is pruning. What most beginners stress over will be over-pruning their plants.

All things considered, the idea of pruning can be challenging to comprehend since the general purpose is by all accounts permitting a plant to develop, not decreasing its mass.

While the idea might appear to be a piece hard to get a handle on, it's quite straightforward. As a matter of fact, different parts of developing plants, such as watering, can utilize comparable rationale to help a plant's general turn of events.

In this aid, we'll make sense of what pruning is, the reason it can help plants, how to go about it the correct way, and a couple of normal errors you can stay away from.

CHAPTER 1: THE OVERVIEW OF PRUNING (EVERYTHING YOU NEED TO KNOW)

PRUNING: what is it ?

Pruning is certainly not a muddled idea. You're scaling back frail and dead material for live plants to advance better development.

At the end of the day, in the event that you have a flower bramble with dead branches, you need to remove those, so your shrubbery invests no additional time into those useless branches.

Consider plants like barrel-shaped compartments with various dig tube lines standing out from the primary body. Associated with those cylinders are more modest empty cylinder lines that have little inflatables appended to more modest bulges.

If you somehow managed to top the primary compartment off with water, the liquid would start to convey equitably all through as you top it off to the top. Each inflatable would start to load up with water.

Nonetheless, with plants, it's not exactly that straightforward. Some of the time the branches (empty cylinders) can become unhealthy or harmed, making the plant squander supplements (water) on ineffective material.

In the meantime, the plant is exhausting extensive energy attempting to take care of a frail, dead, or generally useless branch, which can prompt it not to invest the effort it needs into developing and sustaining sound branches.

It's practically similar to how a languid or problematic representative can cut down efficiency as a director might need to give additional time fixing or rectifying their slip-ups than assisting better specialists with working on their presentation.

All things considered, pruning is a characteristic piece of our general public. We simply call it various things like detaining, terminating, removing, and removing.

REASONS FOR PRUNING

Promote healthy Plant Development

As we examined before, by pruning your plant, you're guiding its energies to advance better, more useful development. This is particularly recognizable with blooming and fruiting plants.

The last thing you need is for your sound branches to seek assets with impasse branches that won't ever create results and undermine your plant's primary honesty. On the off chance that a branch isn't helping your plant, it's stinging it. All aspects of your plant should really buckle down. There's no space for loafers.

Likewise, by lessening the size of your plant, you're opening an upwind stream and offering your better branches more chances to absorb the daylight.

Control Size

You additionally use pruning to control the measures of your plant. Many individuals consider the pruning workmanship known as Bonsai, a pruning procedure that permits individuals to keep smaller-than-expected trees alive once in a while for many years in extraordinary compartments.

In the event that you're growing support or simply attempting to keep your plant at a reasonable size, pruning is what you want to do. This can likewise be applied to developing products of the soil, as you may not need your plant attacking the space of different plants.

Pruning can be an approach to preparing your plants to fill toward the path and to the size you really want them to be.

Train Growth Habit

Certain plants like climbing plants, for example, bounces should be prepared to grow in a specific way. For instance, in the event that you're developing jumps, you show your plant how to develop onto a string and afterward up a post. A similar idea applies to specific tomatoes and establishes the utilization of tomato confines. Also, obviously, this idea applies to fences.

Produce Better Harvest

I discussed how pruning can improve efficiency. This incorporates harvests. For plants that produce natural products, nuts, and veggies, efficiency is estimated by the result of these ending developments.

By pruning, you're controlling your plant so it can zero in its limited energies on developing the most grounded branches which will deliver the best and most heavenly produce.

Further, develop Plant Appearance

Pruning can likewise be utilized to make a specific look. We've all seen those extravagant fences molded like creatures, individuals, and various articles. While your restorative planting objectives probably won't be as aggressive, you could in any case wish to keep things trim and clean.

There's nothing very like strolling through a very much-managed scene. It resembles a split between nature's will to develop and humanity's need to control our current circumstances.

GUIDE TO PRUNING:

Protect people and Property
Pruning can likewise safeguard individuals, pets, and property. Excess of specific plants can harm homes and cause wounds. Envision a flowering hedge that outgrows control, making a way for prickly branches in your yard.

Trees are one of the principal issues with regards to the potential for harming property, as which begins as a bush near your home can rapidly cause issues as branches develop into your windows, siding, drain, and rooftop.

Trees that are not as expected pruned can develop weighty appendages over your property which could cause serious harm. Would it be a good idea for them to decay and fall or sever during a tempest?

Untrimmed trees can likewise give undesirable irritations a pathway into your home.

CHAPTER 2: PRUNING STRATEGIES/TYPE OF PRUNING CUTS

The following are three pruning slices you can use to eliminate an overabundance of foliage or a sick appendage from your plant.

HEADING

With heading, you cut off little parcels of a plant's branches. Doing this permits you to control its size while animating development inside its side stems. Besides, this type of pruning allows you to control the heading your plants develop.

PINCHING

This is a type of pruning that includes the expulsion of your plant's principal stem by squeezing it off with your fingers. This supports the development of new stems while advancing a full plant. Also, doing this assists keep your plant with compacting.

THINNING

Diminishing holds your plant back from getting stuffed by lessening foliage thickness. This type of pruning will take into account better wind stream and more daylight while restricting the spots bugs kind stow away.

PRUNING FRUIT TREES

EARLY SPRING

Organic product trees are normally pruned during late winter, while they are as yet lethargic and their young buds presently can't seem to break. Fences, for example, dogwood and beech, and shrubberies, for example, boxwood and yew are additional instances of plants regularly pruned during the late-winter season. There are additionally a few trees that benefit from pruning during this period.

LATE SPRING

Most elaborate blooming plants and bushes, for example, azaleas and forsythia are pruned soon after their blooms tumble off. Pruning any sooner kills off feasible developing bloom buds.

SUMMER

Evergreen bushes, for example, camellias and Rhododendrons should be pruned during the late spring to be at their hardest for the colder time of year. These shrubberies should be pruned either pre-summer or late-spring as this assists them with keeping up with their shape while eliminating harmed foliage to keep them in the best condition.

FALL

Plants like lavender, gardenias, and explicit kinds of hydrangeas really do best when pruned in the fall while they are going torpid. Along these lines, they can arise throughout the Spring with delightful blossoms.

WINTER

Trees pruned during pre-spring while they are torpid can forestall the spread of illnesses. Vermin, for example, warm-season creepy crawlies search for open tree wounds and are generally dynamic throughout the late spring months.

At the point when such vermin are inert, pruning trees throughout the colder time of year keeps the ailment from spreading.

CHAPTER 3

13 PRUNING ERRORS THAT WILL KILL YOUR NATURAL PRODUCT TREES

What happens when you have a sound organic product tree?

Numerous things!
Not exclusively will the tree make more natural products for your family to appreciate, yet a better natural product tree areas of strength for it opposes nuisance or illness perversions.
Your tree will get by for quite a long time in the future.

Assuming that sounds like what you need for your natural product trees, you should realize what pruning botches you want to stay away from.

I realize I cleared a path with an excessive number of missteps over the years as I figured out how to develop organic product trees! Here are the absolute and generally normal.

1. Investing in the Wrong Tools for Pruning

Having the right apparatuses for pruning natural product trees is significant. Similarly significant is keeping those apparatuses perfect and sterile.

Before you begin pruning your organic product trees, ensure you have the right instruments to get everything taken care of. You should disinfect all instruments before you start pruning to try not to coincidentally move illnesses.

The principal device you want to have is a bow saw.
If you want to eliminate greater appendages, this is the device for the gig.
Search for a general, generally useful pruning saw produced using tempered steel or carbon steel with an agreeable handle that squeezes into your hand.

Then, search for a couple of pruning loppers. Preferably, you need one with adjustable legs that let you arrive at the higher branches without getting on a stepping stool. Loppers eliminate more modest branches or delicate natural product sticks up to 1 inch thick.

The last device you want is a couple of hand pruners to eliminate twigs and delicate natural product sticks under an inch thick. A decent set of hand pruners will last years, so this is speculation. Purchase all that you can manage.

2. Pruning at the Wrong Time of Year

One of the most widely recognized pruning botches that nursery workers make is pruning at some unacceptable season. The best chance to prune a natural product tree is when there are no leaves on the tree, it's lethargic to mean.
Expect to prune your natural product trees in late winter.

Pruning in the spring is gainful because of multiple factors.

In the first place, when there are no leaves on the tree, it's a lot more straightforward so that you might see what you're doing and stay away from botches. Likewise, when you eliminate lethargic buds, it makes the leftover buds begin to quickly develop.

Spring pruning sends the tree into a developing mode.

It is for the most part not prescribed to Prune in the mid-year. At the point when you eliminate such a large number of leaves in the late spring, it eliminates a portion of the food producers for the tree and eases back the fruiting system. It likewise opens natural products to burn from the sun.

The main time that you ought to prune in the late spring is the point at which the tree is Excessively vivaciously developing.
Assuming that occurs, attempt to prune in late spring to give your tree time to mend before it goes lethargic.

3. Forgetting to Prune Every Year

Yearly pruning expands your trees' well-being and efficiency. Put away the opportunity for this significant assignment consistently!

One of the greatest pruning botches you could make is the feeling that you don't have to prune. Certain individuals genuinely think pruning is discretionary - it's not.

Pruning trees consistently is ideal for your tree's well-being and efficiency. It's essential for your obligation when you plant organic product trees on your property. Pruning assists with boosting your tree's well-being and working on the result. You'll wind up with additional organic products when you prune consistently.

4. Not taken off Dead, Dying, and Diseased Branches

Dead, passing on, and illness - the three D's of pruning. Pruning for the three D's is crucial for natural product tree wellbeing!

Some consider this the "Three D's" - dead, kicking the bucket, and sickness - and all nursery workers need to realize this while pruning. Neglecting to eliminate the "Three D's" is no joking matter; you want to eliminate the parts of any green tree in a specific order.

Begin by pruning the dead pieces of the tree first. Doing this eliminates the pieces of the tree that shouldn't get any energy since they won't develop. At the point when you eliminate dead stems, cut three hubs lower down.

Assuming the dead branches were additionally infected, eliminating it guarantees that the sickness doesn't spread. Assuming you see those pieces of the tree that are kicking the bucket, make a point to eliminate them well.

5. Not Following the CAC System

Here is another abbreviation that you can recollect that assists individuals with figuring out how to prune organic product trees appropriately. Everybody needs a system while they're pruning and recall CAC is only that. CAC represents jumbling, intense, and crossing. It gives you a supportive approach to pruning your plants appropriately.

In the first place, eliminate any bunched or messy branches that are viewing for light. These branches take the light from one another, decreasing the tree's capacity to prosper and develop.

Then, eliminate branches at intense points - extremely sharp points. These must be eliminated because they're feeble and won't bear many natural products. Assuming that organic products do show up on these branches, they could snap.

Last, eliminate crossing branches that limit wind stream and advance sicknesses.

Crossing branches is an issue since they rub against one another, causing serious injuries. Wounds give sicknesses a spot to enter your tree.

You could likewise eliminate useless suckers since they're overabundant in development. Your tree shouldn't dedicate energy to suckers.

6. Neglecting to Prune Off Sufficient From The Top

Just removing a few branches and leaves at the highest point of the tree is a typical pruning botch. At the point when you prune trees, the objective is a three-sided shape with fewer branches at the highest point of the tree. It ought to be smaller at the top and more extensive at the base, boosting how much daylight is accessible to your tree.

Assuming you leave an excessive number of leaves at the highest point of the tree, it ingests a lot of daylight than the lower branches. It conceals the lower appendages and causes a mushroom shape to create.

You don't need that!

7. Pruning Off too much

Over-pruning can likewise be an issue. A lot of pruning limits creation and causes organic product trees to battle.

One of the most well-known pruning botches that you make is pruning off a lot from your trees. Over-pruning adds additional pressure to your trees and leads to enduring harm.

Eliminating a lot of your natural product tree removes indispensable buds and diminishes the organic product yield of the tree. Assuming you prune off such a large number of branches, it likewise could make it harder for the tree to ingest the proper measure of daylight.

8. Not Pruning off enough

Then again, under-pruning is more normal than pruning off an excessive amount because individuals stress that, assuming they prune off something over the top, it will kill the tree. Thus, they keep down and just prune off a limited quantity of branches.

You could expect that pruning off a lot is more regrettable than not pruning sufficiently off, but rather that supposition that is off-base. An under-pruned tree conveys a lot of weight on its branches, prompting the branches to snap when the tree proves to be fruitful. Organic product trees develop best when their appendages are short, zeroing in their energy on fruiting and fostering areas of strength for a framework.

9. Allowing Saggy Branches To remain on the Tree

Allow no saggy branches to remain on the tree! You maintain that your branches should develop towards the sun, not the ground. Sagging branches likewise could brush the ground, expanding the gamble of getting a dirt-borne infection.

For what reason do branches become sagging in any case?

Trees send chemicals and supplements streaming vertically all through the tree, causing development and improvement.

Assuming you have sagging branches, it implies that the chemicals and supplements are less plentiful around here.

Branches frequently are less incredible and foster more modest organic products than up-confronting branches. Eliminating sagging branches safeguards against infections while additionally sending more energy toward different pieces of the tree, further developing the natural product quality.

10. Leaving Branches Extremely low on the Fruit Tree

Never leave branches to come up short on your natural product tree. Individuals would rather not cut off branches that have natural products but sit low on the tree. It's not difficult to get up to speed with the number of natural product buds on your tree; you need as many natural products as could reasonably be expected.

Pruning low-hanging branches guarantees you have cleaner foods grown from the ground illnesses. It likewise prompts bigger measured organic products.

11. Not Leaving Spray Access to the Tree

It means a lot to prune to permit admittance to splashes and medicines. Indeed, even natural organic product trees frequently need help with showers and intercessions.

Leaving a mass of branches inside your tree makes it harder for showers to enter your tree. You need a large number of your branches going on a level plane instead of upward. Those branches make it harder for splashes to arrive at each of the regions of your tree.

Many natural products require normal splashing. Try not to expect that you won't have to shower; odds are you will!

12. Eliminating Bigger Branches

Another enormous pruning botch that you would rather not make is pruning huge tree limbs with a width north of 3-4 inches. Assuming you eliminate these branches, it causes wounds that may be excessively enormous for your tree to close, passing on painful injuries for microbes or vermin to enter.

Contingent upon the design of your tree, it's ideal to remove a branch down to the storage compartment, allowing the branch an opportunity to seal.

Another worry while eliminating enormous branches is that they could fall while being pruned. That could harm your tree, severing different branches and tearing bark as it falls. That is the last thing you need!

13. Harming the Tree Husk/ Damaging the Tree Bark

The remainder of the pruning botches you would rather not make is harming the bark. This happens when you prune enormous branches; unintentionally making harm to the bark is conceivable. You never need to tear bark from the side of your trees, harming the tree, and leaving it defenseless against vermin and infections.

Landscapers need to effectively attempt to diminish the gamble of harming tree husks while pruning. Something you ought to do is eliminate more modest branches from the bigger branches to lessen the weight. It lessens the branch falling when you cut it at the collar.

CONCLUSION

As may be obvious, pruning is a vital practice to guarantee your plants develop further and soundly. The primary idea driving it is shedding the frail or dead development from the more prominent plant body to save the better live tissue and the energy the plant commits toward new development like stems, branches, natural products, and blossoms.

The serious issue with committing pruning errors is that it influences the trees into the indefinite future. A few blunders are minor, but if you commit serious errors while pruning, it could harm their natural product creation and well-being for a long time.

Find an opportunity to realize the entirety of the legitimate pruning
strategies before you get everything rolling.

Printed in Great Britain
by Amazon

40021271R00020